Types of Precipitation

水无处不在 降水的分类

[美] 纳迪亚·希金斯 著
[美] 莎拉·英芬特 绘
[美] 艾瑞克·科斯金恩 作曲
沈可君 译

中国水利水电出版社
www.waterpub.com.cn
·北京·

项目策划：徐丽娟
责任编辑：栾　峰　方　斯
特约编辑：李渝汶
联系方式：luanfeng@mwr.gov.cn　010-68545978

书　　名　水无处不在　降水的分类
SHUI WUCHUBUZAI　JIANGSHUI DE FENLEI
作　　者　[美]纳迪亚·希金斯　著　[美]莎拉·英芬特　绘
[美]艾瑞克·科斯金恩　作曲　沈可君　译
出版发行　中国水利水电出版社
（北京市海淀区玉渊潭南路1号D座　100038）
网　址：www.waterpub.com.cn
E-mail：sales@mwr.gov.cn
电　话：（010）68367658（营销中心）
经　　售　北京科水图书销售中心（零售）
电话：（010）88383994、63202643、68545874
全国各地新华书店和相关出版物销售网点
排　　版　陆　云
印　　刷　北京尚唐印刷包装有限公司
规　　格　285mm×210mm　16开本　6印张（总）　80千字（总）
版　　次　2022年1月第1版　2022年1月第1次印刷
总 定 价　148.00元（全4册）

图书在版编目（CIP）数据

水无处不在．降水的分类：汉英对照 /（美）纳迪亚·希金斯著；沈可君译．-- 北京：中国水利水电出版社，2022.1
书名原文：Water All Around Us
ISBN 978-7-5226-0105-2

Ⅰ．①水…　Ⅱ．①纳…　②沈…　Ⅲ．①水－儿童读物－汉、英　Ⅳ．①P33-49

中国版本图书馆CIP数据核字（2021）第210676号

北京市版权局著作权合同登记号：图字 01-2021-5364

亲子学习诀窍

为什么和孩子一起阅读、唱歌这么重要？

每天和孩子一起阅读，可以让孩子的学习更有成效。音乐和歌谣，有着变化丰富的韵律，对孩子来说充满乐趣，也对孩子生活认知和语言学习大有助益。音乐可以非常好地把乐感和阅读能力锻炼有机结合，唱歌可以帮助孩子积累词汇和提高语言能力。而且，在阅读的同时欣赏音乐也是增进亲子感情的好方式。

记住：要每天一起阅读、唱歌哦！

绘本使用指导

1. 唱和读的同时找出每页中的同韵单词，再想想有没有其他同韵单词。
2. 记住简单的押韵词，并且唱出来。这可以培养孩子的综合技能以及英语阅读能力。
3. 最后一页的“读书活动指导”可以帮助家长更好地为孩子讲故事。
4. 跟孩子一起听歌的时候可以把歌词读给孩子听。想一想，音符和歌词里的单词有什么联系？
5. 在路上，在家中，随时都可以唱一唱。扫描每本书的二维码可以听到音乐哦。

扫我听音乐

每天陪孩子读书，是给孩子最好的陪伴。

祝你们读得快乐，唱得开心！

Precipitation happens when water droplets fall from the clouds. It can come in four types. Rain and snow are two forms of precipitation. Sleet happens when rain freezes as it's falling. Hail occurs when rain freezes into balls of ice high up in storm clouds.

Turn the page to learn all about the different types of precipitation.

Remember to sing along!

水滴从云中落下，降水就形成啦。
降水共分四种，
除了雨和雪，
雨点从高空落下的过程中结冰，会形成雨夹雪；
雨滴在高空云层中冻成小冰珠，就形成了冰雹。

请翻到下一页，我们一起来了解降水的不同类型。
跟着音乐一起唱吧！

Rain and snow, sleet and hail,

all water drops from the sky.

雨和雪，雨夹雪，冰雹，
都是水滴，落下云霄。

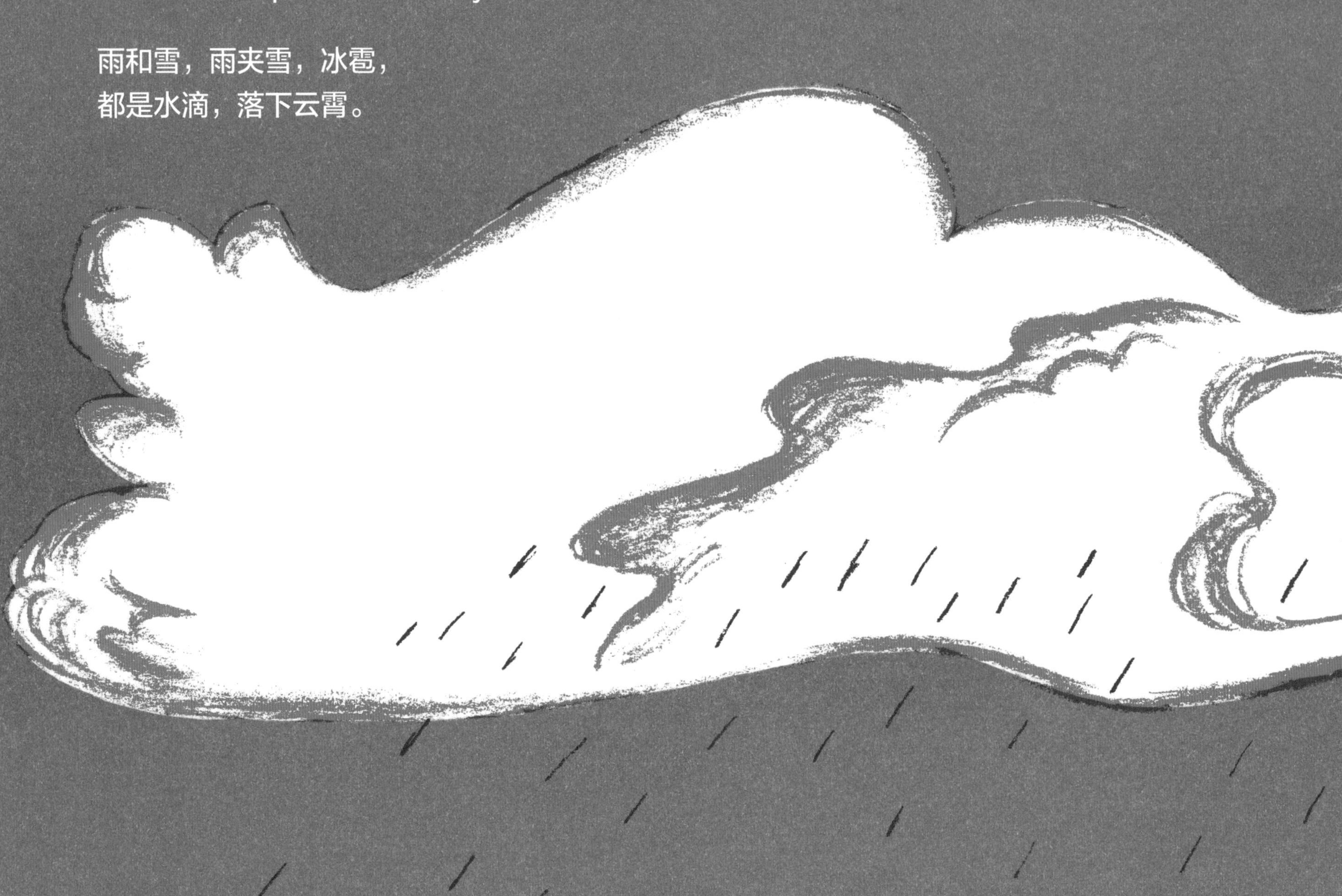

Precipitation comes in four formations
from clouds floating by.

降水不同，分为四种，源自浮云，空中飘动。

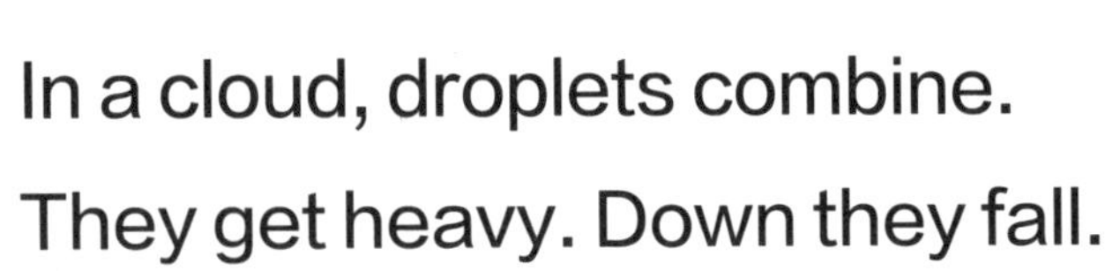

In a cloud, droplets combine.

They get heavy. Down they fall.

小水滴，聚集在云彩里。沉甸甸，凝成雨落下去。

Pitter-patter, pitter-patter.

Rain taps on the wall.

淅淅沥沥，滴滴答答。雨点轻拍，打湿墙瓦。

In a cloud, ice crystals make flakes
so fat they cannot float.

小冰晶，云朵里变雪花，蓬松松，高空中往下洒。

Flutter-glide. Flutter-glide.

Snow falls on the road.

扑扑簌簌，纷纷扬扬。片片雪花，飘散路上。

Rain and snow, sleet and hail,
all water drops from the sky.

雨和雪，雨夹雪，冰雹，
都是水滴，落下云霄。

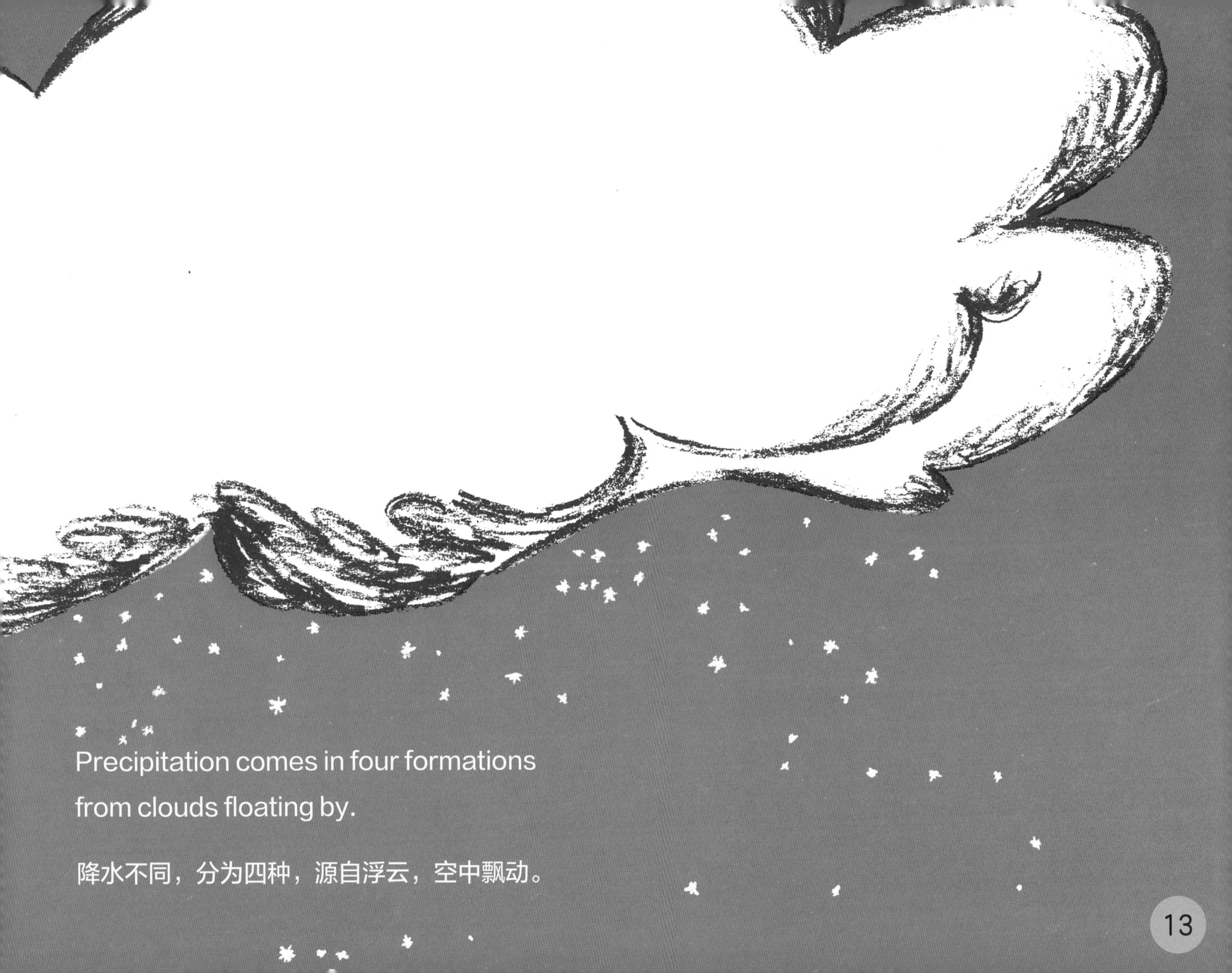

Precipitation comes in four formations
from clouds floating by.

降水不同，分为四种，源自浮云，空中飘动。

From a cloud, wet raindrops fall.

But they freeze on their way down.

小雨滴，从云端落下去。
冷飕飕，在空中结成冰。

Plunkety-plunk. Plunkety-plunk.

Icy sleet hits the ground.

雨水沙沙，雪花飘飘。冰雨夹雪，拍打地上。

In thunder clouds, rain starts to freeze into balls of ice that grow and grow.

雷云里，雨滴结为冰粒。圆溜溜，慢慢涨成冰球。

Crash-bang! Clackety-clang!

Hail knocks on the window.

倏忽砸落，咔啦作响。冰雹袭来，敲打门窗。

Hear the rain splat, rat-a-tat-tat!

Feel the cold snow from your head to your toes.

雨点阵阵，嗒嗒轻叩。
大雪凛凛，浑身颤抖。

See the slick sleet as it bounces on the street.

Make a mad dash when the hail starts to crash.

雨夹雪珠，落地轻跳。
冰雹砸下，快快逃跑！

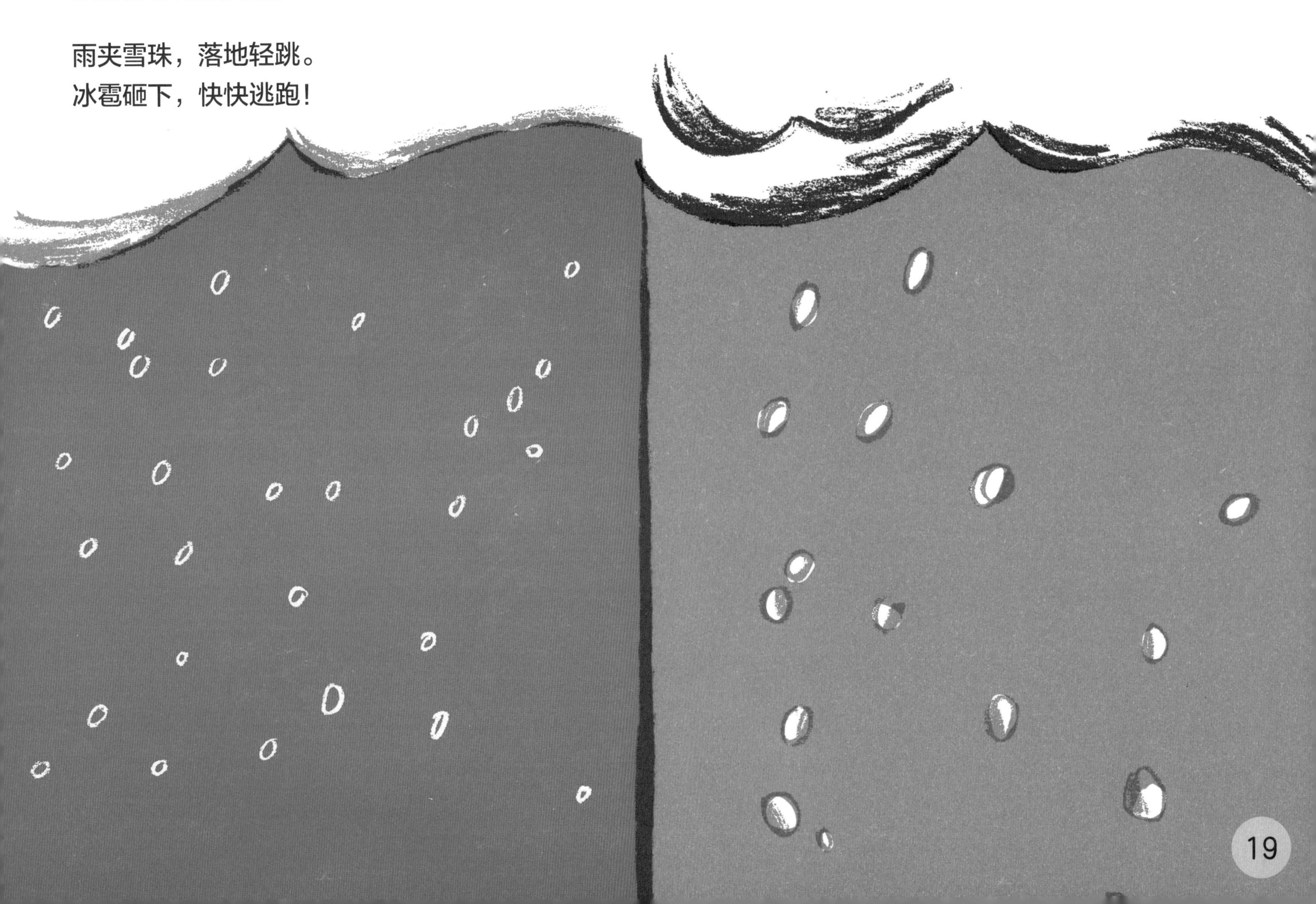

Rain and snow, sleet and hail,
all water drops from the sky.
Precipitation comes in four formations
from clouds floating by.
Rain and snow, sleet and hail,
all water drops from the sky.
Precipitation comes in four formations
from clouds floating by.

雨和雪，雨夹雪，冰雹，
都是水滴，落下云霄。
降水不同，分为四种，
源自浮云，空中飘动。
雨和雪，雨夹雪，冰雹，
都是水滴，落下云霄。
降水不同，分为四种，
源自浮云，空中飘动。

SONG LYRICS 歌词

Types of Precipitation

Rain and snow, sleet and hail,
all water drops from the sky.
Precipitation comes in four formations
from clouds floating by.

In a cloud, droplets combine.
They get heavy. Down they fall.
Pitter-patter, pitter-patter.
Rain taps on the wall.

In a cloud, ice crystals make flakes
so fat they cannot float.
Flutter-glide. Flutter-glide.
Snow falls on the road.

Rain and snow, sleet and hail,
all water drops from the sky.
Precipitation comes in four formations
from clouds floating by.

From a cloud, wet raindrops fall.
But they freeze on their way down.
Plunkety-plunk. Plunkety-plunk.
Icy sleet hits the ground.

In thunder clouds, rain starts to freeze
into balls of ice that grow and grow.
Crash-bang! Clackety-clang!
Hail knocks on the window.

Hear the rain splat, rat-a-tat-tat!
Feel the cold snow from your head to your toes.
See the slick sleet as it bounces on the street.
Make a mad dash when the hail starts to crash.

Rain and snow, sleet and hail,
all water drops from the sky.
Precipitation comes in four formations
from clouds floating by.

Rain and snow, sleet and hail,
all water drops from the sky.
Precipitation comes in four formations
from clouds floating by.

Types of Precipitation

Verse 2
In a cloud, ice crystals make flakes
so fat they cannot float.
Flutter-glide. Flutter-glide.
Snow falls on the road.

Chorus

Verse 3
From a cloud, wet raindrops fall.
But they freeze on their way down.
Plunkety-plunk. Plunkety-plunk.
Icy sleet hits the ground.

Verse 4
In thunder clouds, rain starts to freeze
into balls of ice that grow and grow.
Crash-bang! Clackety-clang!
Hail knocks on the window.

Chorus (x2)

GLOSSARY　词汇表

crystals—glass-like substances that have many sides and are see through

冰晶——像玻璃一样的多面透明体

dash—to run

飞奔——快速奔跑

droplets—very small drops of liquid

小液滴——非常小滴的液体

hail—small balls or lumps of ice that fall from the sky

冰雹——从天空落下的小冰珠或小冰块

sleet—frozen rain

雨夹雪——结冰的雨

读书活动指导

1. 雨、雪、雨夹雪、冰雹是降水的四种不同类型。你最喜欢哪种？为什么呢？
2. 什么季节下雨最多？通常什么时候才会下雪？你见过雨夹雪和冰雹吗？是什么季节见到的呢？
3. 请试着画一幅人们在雨中或雪中的画，他们都穿着什么衣服呢？